LOWER COLUMBIA REVIEW
T 37690
D0371860

Great Ideas of Science

CELL BIOLOGY

by Melissa Stewart

Twenty-First Century Books
Minneapolis

For Mom, who spent many hours looking at cells under a microscope

Twenty-First Century Books
A division of Lerner Publishing Group, Inc.
241 First Avenue North
Minneapolis, MN 55401 U.S.A.

Website address: www.lernerbooks.com

Library of Congress Cataloging-in-Publication Data

Stewart, Melissa.
Cell biology / by Melissa Stewart.
p. cm. — (Great ideas of science)
Includes bibliographical references and index.
ISBN-13: 978–0–8225–6603–8 (lib. bdg. : alk. paper)
1. Cytology—Juvenile literature. I. Title.
QH581.2.S74 2008
571.6—dc22 2006028542

Manufactured in the United States of America
1 2 3 4 5 6 – DP – 13 12 11 10 09 08

TABLE OF CONTENTS

INTRODUCTION

COLOSSAL CELLS

As the sun's early-morning rays filter across an African plain, an exhausted leopard scrambles up a tree trunk. The big cat stretches out on the lowest limb and drifts off to sleep. Moments later a hulking hippo passes by, following a well-worn trail back to its favorite water hole. After a night of feeding, the hippo is ready to lounge in the shallow water. Off in the distance, zebras and gazelles graze skittishly. They are always on the lookout for danger.

But it's an alert male ostrich that first spies trouble: four spotted hyenas in search of a morning meal. The bird's height gives it a clear advantage as it scans the savanna for predators.

The male ostrich takes off at top speed, fleeing at 40 miles (65 kilometers) per hour. The zebras and gazelles run too. But the female ostrich must stay put and protect her eggs. She crouches close to the ground, stretches out her long neck, and remains perfectly still. From a distance, she looks like a small bush. The hungry hyenas trot off in another direction.

Ostrich eggs are much larger than chicken eggs. Ostrich eggs are the largest eggs on Earth, and they contain some of the largest cells.

Most mother birds fiercely protect their eggs, but ostrich eggs are particularly hard to hide. The sturdy, football-sized shells contain as much yolk and white as two dozen chicken eggs. Not only are they the largest eggs on Earth, but ostrich eggs also contain some of the largest cells.

A cell is the basic unit of life. It is the smallest structure that can carry out all the processes necessary to sustain life, from turning food into energy to creating offspring. The yolk cell in an ostrich egg is the size of a baseball, making it the world's largest cell by volume. The longest cells in the world are found in giraffes. The nerve cells in a giraffe's neck can be nearly 10 feet (3 meters) long.

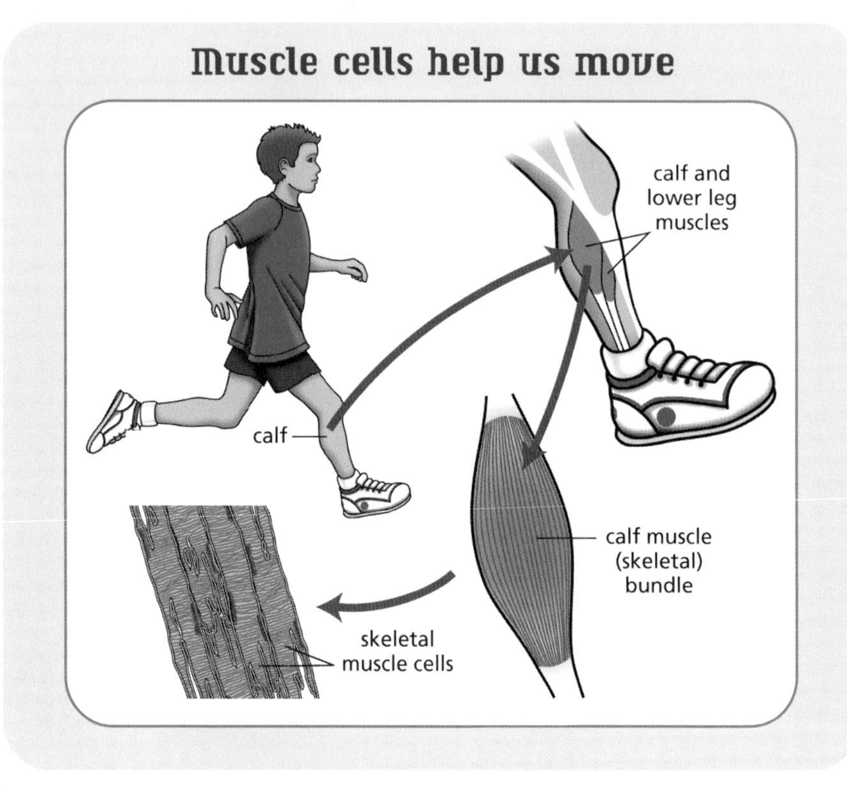

But these colossal cells are notable exceptions. Most cells are so tiny that you need a microscope to see them. A typical cell in the human body is about 1/1,000th of an inch (0.0025 centimeters) wide. It would take nearly fifty of those cells to stretch across the period at the end of this sentence.

Your body contains about 100 trillion cells—muscle cells, nerve cells, blood cells, and more. Each type of cell is designed to do a specific job. Red blood cells look like tiny saucers. As they circulate through your body, they deliver oxygen and pick up carbon dioxide. Muscle cells are long, thin, and stretchy. They help you run across a soccer field and raise your hand in class. Nerve cells look something like a spiderweb. They carry messages to and from your brain.

Most of the cells in your body do not work alone. Large numbers of similar cells clump together to form a tissue. All the cells in a tissue work as a team. Groups of different tissues join together to form even larger structures, called organs. The heart is an organ that contains blood tissue and muscle tissue. Your stomach, lungs, and

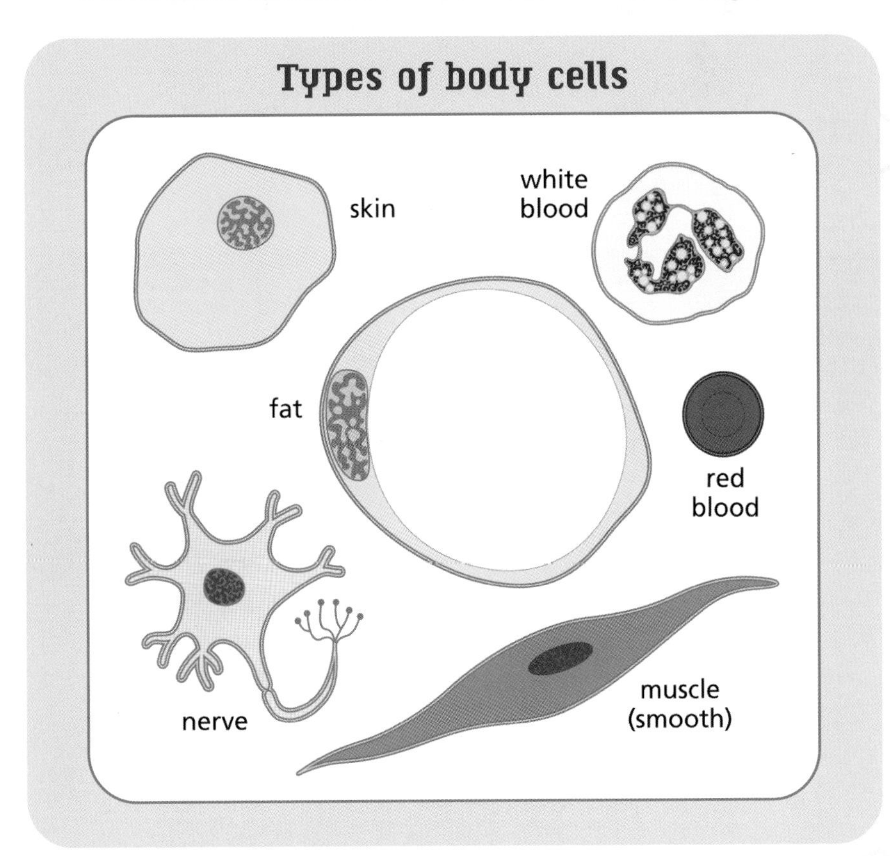

brain are organs too. They each contain different combinations of tissues.

Each of your organs is part of an organ system. Your stomach is an important part of your digestive system. The stomach works with the esophagus, small intestines, large intestines, pancreas, and gallbladder to break down food and move nutrients into your blood. Other cells, tissues, and organs make up your nervous system, immune system, respiratory system, and circulatory system. Each system plays a unique role in keeping you alive and well. When all your organ systems function normally and cooperate with one another, you can live and grow.

Some tiny living things consist of just one cell. If you look at a drop of pond water under a light microscope,

Freshwater organisms found in a drop of pond water

you might see dozens of these tiny, single-celled creatures. You can also find them in sand, soil, and just about every other surface on Earth. For these creatures to survive, their single cell must do many different jobs. It must take in nutrients, get rid of wastes, avoid enemies, and produce offspring.

Cell biology is the study of all cells—from chicken eggs and human skin cells to single-celled amoebas and the cells that make up a rose petal. Whether a cell is as large as an ostrich egg's yolk or as small as a bacterium, each and every one is a marvel of design and efficiency. Cells carry out thousands of biochemical reactions every minute and reproduce new cells to keep life going.

This book tells the story of cell biology. It describes how the field developed—sometimes slowly, sometimes by leaps and bounds. Because cells are so small, no one even knew they existed until microscopes were invented. Each time scientists devised new equipment and techniques for viewing cells, the field exploded with discoveries. Then research stalled again—until scientists found another new way to study the basic units of life.

As knowledge of cells grew, some scientists decided to dedicate their whole careers to studying just one kind of cell. Others focused their research on one part of the cell or a specific cell function. In this way, the field of cell biology has slowly expanded, giving birth to many new branches of biology.

Chapter 1

The Magic of Microscopes

Four hundred years ago, no one knew that an egg yolk was a giant cell. In fact, people didn't even know that cells exist. But by the end of the 1600s, thanks to a new invention, a few curious people had begun to observe the microworld—an amazing array of creatures invisible to the naked eye. What was that incredible new invention? The microscope.

In the late 1500s, the Netherlands emerged as the world leader in making lenses for eyeglasses. After spending the workday crafting eyeglasses, Dutch lens makers experimented with new designs and looked for different ways to use their lenses.

During their late-night tinkering, Hans Janssen and his son Zacharias found that when they placed two lenses at just the right distance from one another, they created a magnified (larger) image of an object. Using this knowledge, the father-and-son team designed and built the world's first compound, or multi-lens, microscopes. Each consisted of a tube with a lens at both ends. The devices could magnify images up to ten times larger than life size.

Cells: An Exciting Discovery

Building on the Janssens' basic design, lens makers in the Netherlands, Italy, and Great Britain improved the compound microscope. Within fifty years, microscopes could magnify objects up to thirty times their actual size. That's when the devices caught the attention of British scientist Robert Hooke.

Robert Hooke

Hooke has been described as the greatest experimental scientist of his time. His interests ranged from physics and astronomy to chemistry, biology, and geology. He worked as a surveyor and architect in London and developed new forms of naval technology. Some of Hooke's inventions are still used in cameras, automobiles, and wristwatches. Perhaps most important, he worked with an instrument maker named Christopher Cock to create the first microscope with a built-in light source. This made microscopes much easier to use.

Hooke used his microscope to examine sand, sugar,

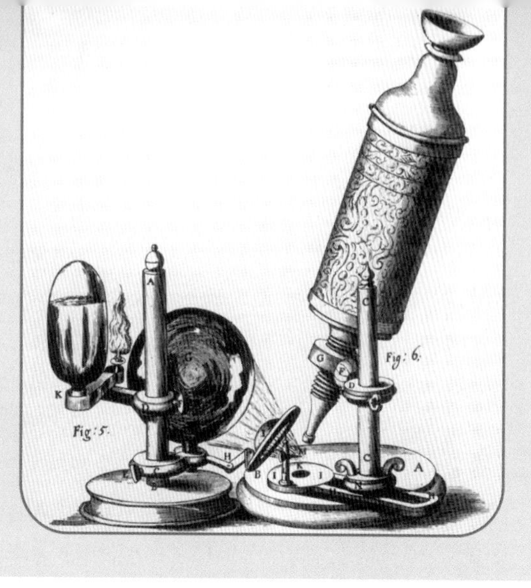

This drawing shows an early microscope designed by Robert Hooke in 1665.

hair, tears, insects, bird feathers, snowflakes, fossils, and plants. When he looked at a slice of cork, he was amazed by what he saw: dozens of small, empty spaces surrounded by walls. These mysterious structures reminded Hooke of the small, plain rooms in a monastery or a jail, so he called them "cells."

Cork is not alive, but it is made from a type of oak tree that grows near the Mediterranean Sea. Hooke was the first person to see dead plant cells. In 1665 he described his observations in a book called *Micrographia*. It contained many drawings, including one of cork cells.

Other Early Observers

Around the same time, Marcello Malpighi, an Italian scientist and doctor, was studying the internal structure of plants and animals. One of his favorite subjects was frogs. Malpighi made thin slices of frog body tissues and examined the brain, liver, kidney, spleen, lungs, bones, and skin with a compound microscope. Malpighi discovered the taste buds in tongue tissue. He even may have understood their role in sensing food particles.

While viewing frog lungs in 1666, Malpighi noticed a network of tiny blood vessels. He named them capillaries. Malpighi correctly suggested that capillaries connect arteries to veins and allow blood to flow back to the heart.

In 1671 Malpighi published a book that included diagrams of a chick's early development. This work was so important that many scientists consider Malpighi the "father" of developmental biology. This branch of science, sometimes called embryology, focuses on embryo growth and development.

Later Malpighi examined plants in great detail. He noticed that plant leaves are made of tiny units. Malpighi didn't know that Hooke had already named these structures *cells*. Malpighi called them *utricles*, which means "small sacs" in Latin. Malpighi was probably the first scientist to identify living cells. Additional studies of plant roots and leaves convinced him that all plant tissues were made up of cells. He claimed that each cavity was filled with fluid and surrounded by a firm cell wall.

In England a physician named Nehemiah Grew was also studying plant cells. Like Malpighi, Grew recognized that cells are the basic structural units of plant tissue. While examining plant leaves in 1682, Grew noticed large numbers of small green structures inside each cell. Grew had spotted chloroplasts. They collect the sunlight that plants need to make their own food. Scientists later learned that all cells contain several different kinds of structures, called organelles. Each organelle does a specific job inside the cell.

The Compound Microscope: A Closer Look

Even a quick glance at a compound microscope will tell you it's a complicated device. But a microscope is easy to use once you learn the names and jobs of its major parts.

A compound microscope

Parts of a Compound Microscope

Name	**Function**
Eyepiece	**Has a lens that magnifies objects 10 times**
Body tube	**Supports the eyepiece**
Arm	**Houses the focusing knobs**
Stage	**Holds specimens (samples of materials)**
Nosepiece	**A rotating wheel with three "objective" lenses—lenses used to view objects. The lenses usually have magnifications of 10 times, 40 times, and 100 times.**
Base	**Contains light source**

When you look at a specimen, such as body tissue, with the shortest objective lens—the one marked 10—the image is magnified 100 times (10 x 10). When you use the medium-length objective lens, the total magnification is 400 times (10 x 40). With the longest objective lens, the total magnification is 1,000 times (10 x 100).

Master of the Simple Microscope

Antoni van Leeuwenhoek

Meanwhile, a Dutch cloth merchant named Antoni van Leeuwenhoek was developing a different kind of microscope. Leeuwenhoek often used a hand lens, or magnifying glass, to inspect the quality of cloth before he bought it. Perhaps because he wanted a better magnifier than his competitors had or perhaps because he was inspired by Hooke's *Micrographia*, by the 1670s Leeuwenhoek was spending much of his free time building and using simple microscopes (microscopes with a single lens).

Leeuwenhoek was not trained as a scientist, but he was curious, patient, hardworking, and open to new ideas. These personality traits helped him excel at making and using lenses. Leeuwenhoek also was a careful observer, and he kept detailed notes about everything he saw. He couldn't draw well, so he hired artists to create illustrations of the things he observed.

In 1673 Leeuwenhoek began writing long, detailed letters about his work to the Royal Society of London, a group of respected scientists in Great Britain. The society members were impressed by Leeuwenhoek's work. They published many of his letters and drawings.

In one letter, Leeuwenhoek reported what he saw in a sample of saliva from his own mouth: "many very little, living animalcules, very prettily a-moving." When he examined an old man's saliva, Leeuwenhoek observed "an unbelievably great company of living animalcules. . . . The biggest . . . bent their bodies into curves in going forwards . . . The other animalcules were in such enormous numbers, that all the water . . . seemed to be alive."

What exactly were the animalcules, or little animals, that Leeuwenhoek described? Modern scientists believe that they were tiny, single-celled organisms, eventually known as bacteria. Even though most bacteria are about ten times smaller than animal cells, Leeuwenhoek was able to see them.

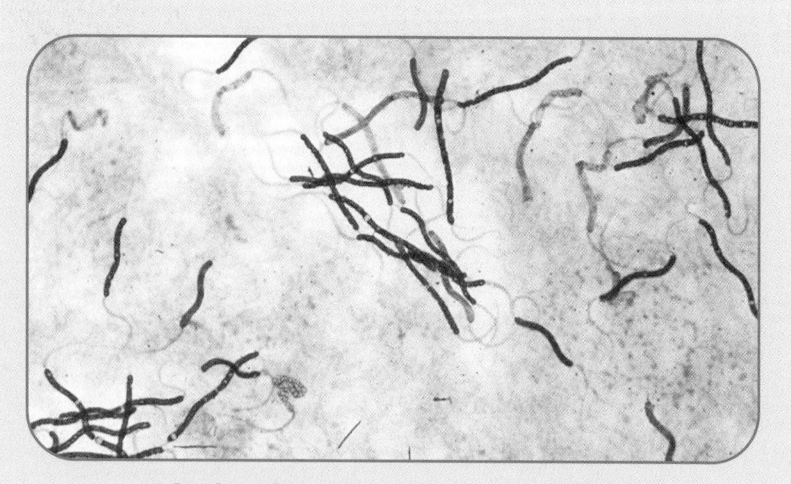

A look at bacteria under a microscope

Over the next fifty years, Leeuwenhoek spent countless hours studying the muscles, nerves, hair, bones, skin, and eyes of many animals. He was the first person to describe red blood cells and sperm cells. He also observed a group of microorganisms called protists. Protists are larger and more complex than bacteria. Scientists believe that protists gave rise to all fungi, plants, and animals.

Leeuwenhoek's Microscopes

Leeuwenhoek's microscopes didn't look much like modern microscopes. Each one was just a few inches long and fit easily in his palm. A tiny lens fit snugly into holes in two brass plates that were riveted (bolted) together.

Leeuwenhoek stuck solid specimens onto a pin below the lens. He poured liquid specimens into a small tube that could be glued to the pin. Using a screw attached to the pin, Leeuwenhoek could rotate the specimen. He could also move it back and forth. This is how he brought objects into focus.

Leeuwenhoek's microscope could magnify objects to more than two hundred times their actual size.

The position of the lens depended on the specimen's size and shape. Because each one was different, Leeuwenhoek needed to craft a new microscope almost every time he wanted to look at a specimen. During his lifetime, Leeuwenhoek built at least 250 microscopes.

Viewing an object through one of Leeuwenhoek's simple microscopes wasn't easy. He had to hold the device very close to his eye. Focusing the microscope took a lot of patience, but his effort was rewarded. Leeuwenhoek's microscopes were much better than any others available at the time. Most compound microscopes in the late 1600s could magnify objects forty or fifty times. Leeuwenhoek's microscopes magnified objects to more than two hundred times their actual size. The resolution, or ability to see details clearly, was also far better.

Impressed by Leeuwenhoek's microscopes, lens makers began attaching a simple microscope to the back of a compound microscope. These devices remained popular until the early 1800s, when new technologies made it possible to build better compound microscopes.

CHAPTER 2

A CLOSER LOOK AT PLANT CELLS

During the 1700s, discoveries based on microscopic observations declined, due to the limitations of lenses. Lens makers and scientists struggled to make objects look less blurry at high magnifications. They also searched for ways to reduce halos—hazy streaks of color that made viewing difficult.

In the 1730s, Chester Moor Hall, a British lawyer who enjoyed experimenting with lenses in his free time, noticed that if he placed two different kinds of glass close together, no halos appeared. The second piece of glass canceled out any defects in the first piece.

Hall didn't realize the importance of his discovery, but a telescope maker named John Dollond did. In 1759 Dollond tried to use Hall's design to build microscopes. But he found that making a combined-lens system small enough for a microscope was very difficult. Dollond never figured out a way to make Hall's idea practical.

A breakthrough finally came in the early 1800s, when a Dutch father and son, Jan and Harmanus van Deyl, developed microscope lenses that did not produce annoying colored halos. Their design gave scientists a clear view of small objects. The new microscopes still did not have enough magnifying power to see the details of animal cells. But they were perfect for viewing plant cells, which are about three times larger than animal cells.

Moving in and out of Cells

Around this time, a French researcher named Charles-François Brisseau de Mirbel made a bold claim. Based on microscopic studies of primitive plants called liverworts, Mirbel proposed that all plants are composed entirely of cells. As researchers began to embrace Mirbel's idea, they tried to explain how plant cells form. Although most of their ideas were later disproved, these scientists did contribute to knowledge of the cell. Hugo von Mohl, a plant scientist in Germany, described several different kinds of plant tissue. Two other Germans, D. H. F. Link and Karl Asmus Rudolph, showed that all plant cells have their own walls. And French researcher Henri Dutrochet discovered how some materials flow in and out of cells.

One crucial substance that moves in and out of cells is water. A healthy human body is made of 55 to 65 percent water. It's inside your cells and all around them. For a cell to be stable, the amount of water inside and outside the cell must be equal. If the concentration of water is not the same on both sides of the cell membrane—the layer of molecules that surrounds and encloses the cell—water moves from the area where its concentration is higher (there is more water) to the area where its concentration is lower.

Why Are Cells So Small?

When a cell is small, the surface area of its membrane is large compared to how much material is inside. Organelles lie fairly close to the cell membrane, and nutrients, minerals, and other supplies find plenty of spaces to slip through the cell membrane. Organelles get the raw materials they need to do their jobs and have no trouble getting rid of their waste products.

As a cell grows, the volume of material inside the cell grows faster than the surface area of its membrane. Larger cells carry out more chemical activities than smaller cells, producing more wastes. Their organelles need more raw materials, and wastes must be removed more quickly. But in larger cells, many organelles are farther from the cell membrane. Transporting supplies to the organelles takes more time, and shuttling wastes away requires more effort.

Because the cell membrane's surface area hasn't grown much, there are only a few more places for supplies to enter the cell and wastes to exit. Organelles have a harder time getting the nutrients, minerals, and other materials they need. At the same time, waste products start to pile up. The cell cannot function as well.

In this way, a cell's size is limited by the relationship between the surface area of its membrane and the volume of its insides. A small cell has enough surface area to meet all its needs, but a large one may not. Most of the cells in your body are about 1/1,000th of an inch wide. If they were any larger, they would not be able to function.

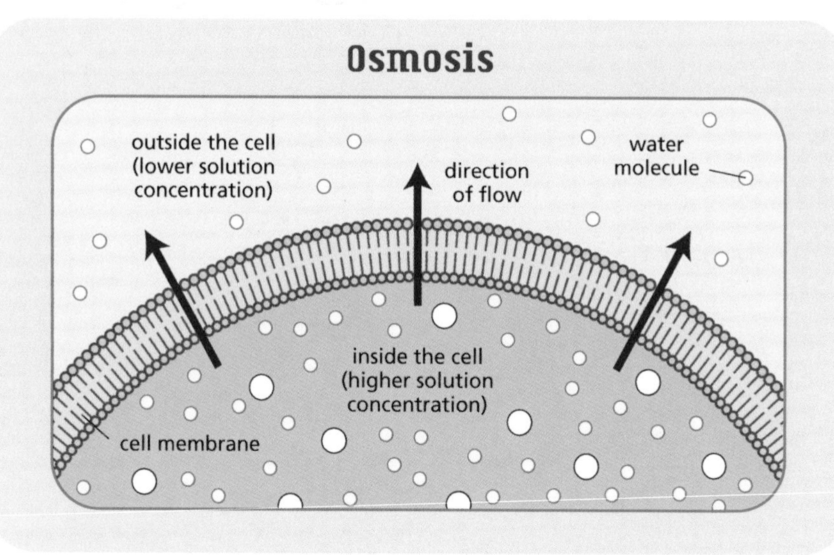

This process is called osmosis.

Water isn't the only material that travels in and out of cells. Everything your cells need to survive, from carbohydrates and proteins to vitamins and minerals, must pass through the cell membrane. As chemical reactions occur inside a cell, they produce waste. The waste materials leave the cell through the cell membrane.

Different materials enter and exit the cell in different ways. Some food particles and ions (electrically charged atoms) move in and out of the cell the way water does. When their concentration is low inside the cell, they flow in. When their concentration is high inside the cell, they flow out. This movement is called diffusion. (Osmosis is a type of diffusion involving water.)

In 1827 another French scientist, François-Vincent Raspail, suggested that some large molecules enter cells

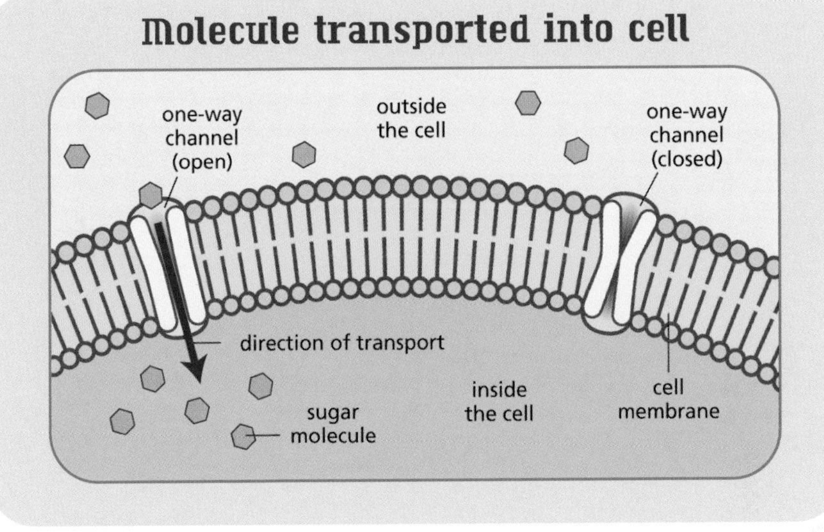

through special one-way channels in the cell membrane. Raspail called this process selective transport. These large molecules can move into cells even if their concentration is greater inside than outside. This allows cells to stockpile proteins, sugars, minerals, and other important substances. By building up an extra supply of nutrients, cells can prepare for an emergency. For example, stockpiled sugar molecules can be used to produce a boost of energy at a moment's notice.

Studying Stomata

While François-Vincent Raspail and Henri Dutrochet were studying how materials move across the cell membrane, a German researcher named C. L. Treviranus was observing stomata—tiny structures on the underside of plant leaves. Treviranus's work united what scientists knew about plant cell functioning with what they knew about plant anatomy, (the study of the parts or structure of an organism).

Stomata are simple structures. Each stoma consists of two guard cells that can contract (squeeze) or relax. When the guard cells are relaxed, a small hole opens to the outside world.

Treviranus proposed that guard cells control the flow of two gases—oxygen and carbon dioxide—in and out of leaves. Treviranus based his idea on the work of Swiss physicist Nicolas-Théodore de Saussure.

In 1804 Saussure had become intrigued by a famous experiment conducted by Flemish doctor Johannes (Jan) Baptista van Helmont in the 1600s. Van Helmont wanted to know if plants get the energy they need to live and grow from nutrients in soil. To find out, he planted a young willow tree in a large bucket and monitored the tree's growth.

After 5 years, the tree had gained 164 pounds (74 kilograms). But the soil was only 2 ounces (62 grams) lighter. Clearly, soil was not the main source of the tree's nutrients. Van Helmont decided that most of the nutrients must come from water, but he never tested this hypothesis. A few years later, Stephen Hale, a British clergyman and scientist, suggested that plants might also absorb nutrients from the air.

Hoping to settle the matter, Saussure repeated Van Helmont's experiment. As the willow tree grew, Saussure carefully measured the amount of water he gave it. He also measured the gases the tree took in and gave off. His results showed that to live and grow, a plant needs water and carbon dioxide. At the same time, it releases oxygen.

With Saussure's results in mind, Treviranus proposed that a plant leaf takes in carbon dioxide and releases oxygen through its stomata. Despite this important discovery, Treviranus could not figure out how plant cells convert water and carbon dioxide into energy.

The Basics of Photosynthesis

In 1842 Julius Robert Mayer, a German physician and physicist, finally put together all the pieces of the puzzle. Mayer was familiar with the work of two earlier scientists, Joseph Priestley, a chemist from Great Britain, and Jan Ingenhousz, a physician-physicist from the Netherlands. In the early 1770s, Priestley had shown that plants give off oxygen only when exposed to sunlight. A few years later, Ingenhousz discovered that only the green parts of plants—the stems and leaves—release oxygen.

Armed with this information, Treviranus's findings, and a general knowledge of physics, Mayer came up with a theory. He suggested that when enough water and carbon dioxide are available, the cells in a plant's leaves and stems can convert light energy from the sun into chemical energy.

Joseph Priestley

This chemical reaction produces oxygen. The whole process is called photosynthesis.

Mayer believed that plant cells used the chemical energy created during photosynthesis to fuel their activities. Scientists later learned that photosynthesis is more complex. Plant cells can't use chemical energy directly. Instead, they use it to fuel a series of chemical reactions that produce a kind of sugar called glucose. Glucose gives plant cells the energy they need to live and grow.

Even though photosynthesis is a complicated process, it can be represented by a fairly simple chemical equation:

$$6\ CO_2 + 6\ H_2O + \text{sunlight} \rightarrow C_6H_{12}O_6 + 6\ O_2$$

According to this equation, when six molecules of carbon dioxide (CO_2) combine with six molecules of water (H_2O) in the presence of sunlight, the result is one molecule of glucose ($C_6H_{12}O_6$) and six molecules of oxygen (O_2).

In the 1860s, a young German scientist took Mayer's theory a step further. Julius von Sachs was interested in the structure and function of plant cells. He was especially intrigued by the work two French chemists, Pierre-Joseph Pelletier and Joseph Caventou, had done in 1817. By soaking crushed plant leaves in a series of solvents, the researchers had isolated the pigment that gives leaves their green color. They named the substance chlorophyll.

In 1865 Sachs discovered that chlorophyll isn't spread evenly throughout plant cells. It's located only in tiny organelles called chloroplasts. Chloroplasts were first described by Nehemiah Grew in the 1670s, but it was Sachs who determined their role in the cell. Chloroplasts make glucose during photosynthesis.

The Trouble with Animal Cells

Scientists were finally beginning to understand how plant cells function. But research on animal cells lagged behind. Because animal cells are much smaller than plant cells, microscopes were not powerful enough to observe them. Researchers also found it difficult to prepare animal samples for viewing.

A compound microscope can create an image of an object only if light can pass through the object. Because plant leaves are thin and flat, it is easy to cut them up and place a small sample on a slide. It's also fairly easy to cut a thin section of a plant's stem or root with a sharp razor blade. But slicing up animal tissue is much more time-consuming and messy.

Both of these problems were solved in the early 1830s. First, British wine merchant and entrepreneur Joseph Jackson Lister designed a microscope that could magnify objects 400 times with detailed resolution. A few years later, Czech scientist Jan Purkinje and German scientist Gabriel Valentin invented the microtome—a device that cuts thin, even slices of animal tissue. About the same time, another German researcher, Robert Remak, developed a hardening agent that made animal tissues firm, so they were easier to slice. In less than a decade, these powerful new tools revolutionized the study of cells.

CHAPTER 3

THE CELL THEORY TAKES SHAPE

Scottish scientist Robert Brown had collected about four thousand plant species while traveling in Australia. Using a microscope based on Joseph Jackson Lister's design, Brown spent countless hours studying the plants. In 1833 he noticed tiny "opaque spots" near the center of cells in orchid stems. When Brown looked at cells from other parts of the plant, he saw the same strange spots.

Intrigued by his discovery, Brown looked for similar spots in cells from other plants. When he found them, he decided that they must be a feature of all plant cells. He named the spots "nuclei." The word is the plural of *nucleus*, a Latin term meaning "kernel" or "central part."

Brown guessed that nuclei were important, but he wasn't sure what their purpose might be. At the time, most botanists (scientists who study plants) focused on naming and describing plants. Brown set aside his work on nuclei and continued to study his huge collection of Australian plants.

Cells: The Basic Unit of Life

When German botanist Matthias Jacob Schleiden learned of Brown's work, he was fascinated. Schleiden decided to take a closer look at cell nuclei. He agreed with Brown that nuclei must play an important role in cell functioning. By 1838 Schleiden was also convinced that plants grow by producing new cells.

Schleiden mentioned his ideas to German scientist Theodor Schwann at a dinner party. As Schleiden described his observations, Schwann was stunned. He realized that plant cells had much in common with the animal cells he was studying. If nuclei were as important as Schleiden claimed, animal cells should also have them. Schwann invited Schleiden to visit his lab at the University of Leuven in Belgium.

Working together, Theodor Schwann *(left)* and Matthias Jacob Schleiden *(right)* discovered that cells are the building blocks of life.

When the two scientists looked closely at animal tissues, they spotted tiny nuclei in almost every cell. Schleiden and Schwann concluded that cells are the basic units of both plants and animals. Schwann described this revolutionary idea, known as the cell theory, in a paper published in 1839. Schwann wrote, "We have seen that all organisms are composed of essentially like parts, namely of cells."

Most groundbreaking research goes through a period of debate in the scientific community, but the cell theory gained immediate acceptance. Following the publication of Schwann's paper, many other scientists came up with even more evidence to support the theory.

How Cells Reproduce

Developing the cell theory was a major accomplishment. But as scientists continued to study cells, they discovered that not all of Schleiden and Schwann's ideas were quite right. For example, Schleiden and Schwann believed that new cells form spontaneously.

The idea of "spontaneous generation" dated back to ancient Greece. For more than two thousand years, people had believed that some living things arise spontaneously from nonliving matter. According to a recipe from the 1600s, mice could be created by placing sweaty underwear and wheat husks in an open jar for three weeks. The sweat from the underwear was believed to enter the wheat husks and change them into mice.

The Road to a Great Idea

When Schleiden and Schwann's ideas became popular, some other scientists protested. They thought that they deserved some or even all of the credit for developing the cell theory. And they had a point. Important ideas such as the cell theory—one of the central principles of modern biology—do not usually appear out of the blue. The evidence that led to the cell theory had been slowly accumulating since the 1660s, when Robert Hooke first saw cells.

As early as 1805, a German naturalist-philosopher named Lorenz Oken had proposed that both plants and animals are made up of "infusoria." This Latin word is used in modern times to describe tiny aquatic creatures, but Oken probably used it more broadly to refer to cell-like units. He had no way to prove his claim, but his ideas were not forgotten.

In the 1820s, Raspail and Dutrochet both suggested that cells are the basic units of plants and animals. And in the early 1830s, Purkinje and Valentin observed that animal tissues, like plant tissues, are composed of cells.

German scientists Johannes Müller and Friedrich Henle were also studying cells at the time. Like other researchers, they recognized that cells are the basic building blocks of both plants and animals. But it was Schleiden and Schwann who clearly and boldly asserted, "There is one universal principle of development for the elementary parts of organisms, however different, and this principle is the formation of cells."

The first serious attack on spontaneous generation came in 1668. At the time, most people believed that maggots formed spontaneously in rotting meat. But Francesco Redi, an Italian doctor and poet, suspected that maggots came from eggs laid by flies. To test his hypothesis, Redi placed pieces of rotting meat in three different types of containers. Some containers were tightly sealed, some were open to the air, and some were covered with a cloth. Maggots appeared only in the open containers, where flies could have landed and laid eggs.

Despite Redi's results, most people continued to believe that new creatures formed spontaneously. Early microscopic observations seemed to back up this idea. Scientists saw a whole new world of tiny organisms that appeared to arise from nothing.

In 1745 John Needham, an English clergyman, proposed an experiment to test the idea. He boiled chicken broth, poured it into a glass container, sealed it, and waited. Within a few days, microorganisms appeared. According to Needham, this proved that spontaneous generation does occur.

But an Italian priest named Lazzaro Spallanzani was not convinced. He thought the microorganisms might have entered the broth after it had been boiled, but before it was sealed. To test his idea, he poured chicken broth into a glass container, sealed it, and removed the air before heating the contents. To Spallanzani's delight, no microorganisms grew. But critics said his results proved only that spontaneous generation could not occur without air.

Given the apparent evidence, Schleiden and Schwann thought that cells form by spontaneous generation. If it could work for whole organisms, then why shouldn't it work for their basic units?

But in the early 1850s, Belgian botanist Barthélemy Dumortier reported observing plant cells splitting in half to create new cells. Around the same time, Robert Remak observed the same phenomenon in animal cells. But it was Polish scientist Rudolph Virchow who expressed this new idea best, declaring, "All cells come from cells."

Virchow's landmark scientific paper was published in 1858. He showed that new cells form when existing cells divide in two.

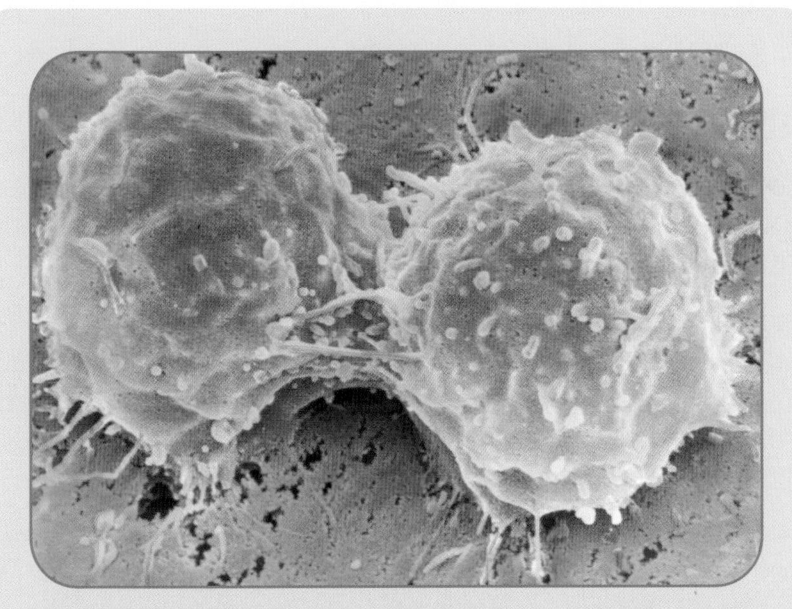

This photograph shows a nerve tissue cell dividing into two new cells.

By 1860 scientists had firmly established the three major principles of cell theory:

1. All organisms consist of one or more cells.
2. The cell is the basic structural and functional unit of all organisms.
3. All cells arise only from preexisting cells.

While future generations of scientists would learn much more about cell structure and function, cell theory's basic ideas have remained unchanged. Modern researchers consider it one of the most important concepts in biology. Cell theory shows that all creatures on Earth—no matter how simple or how complex—have something very important in common. They are all made of the same basic units or building blocks.

CANCER CELLS **Most of the cells in your body are constantly growing. When they reach their size limit, they split in half to form two new cells. Cell division makes it possible for single-celled creatures to create a new generation. In plants and animals, new cells fuel growth and development.**

But cells don't live forever. Most human cells are programmed to divide about fifty times. Then they die. Once in a while, cells do not follow their built-in instructions. They grow and divide very quickly, and they never stop. This uncontrolled cell reproduction can cause a disease called cancer.

In some cases, doctors can remove cancer cells during surgery. In other cases, treatment with radiation, drugs, or both destroys cancer cells. But sometimes doctors cannot stop the cancer.

CHAPTER 4

INSIDE THE NUCLEUS

Scientific knowledge often builds slowly and steadily over long periods of time. But sometimes science advances in leaps and bounds, sputters to a halt, then lurches forward again. This is how researchers' understanding of cell biology grew.

Advances in microscopes and techniques for viewing cells fueled the sudden surge of activity in the 1830s that led to Schleiden and Schwann's cell theory. A similar wave of discovery came in the 1880s and 1890s. Once again, rapid progress happened as a result of improved viewing techniques. The work that led up to those improvements began in the 1850s.

Scientists working separately in different parts of Europe stumbled onto the same discovery at around the same time. They realized that staining cells with dyes made specific cell structures stand out. The contrast between the colorful dye and the colorless background made the cell parts much easier to see.

Throughout the 1860s and 1870s, scientists experimented with a wide variety of staining techniques. They found that different types of stains, or dyes, added color to different cells or cell parts. For example, methylene blue stain highlights cell nuclei, while iodine stain has the right chemical properties to stick to carbohydrates, making them appear blue. A dye known as Wright's stain makes red blood cells pink, while crystal violet stain turns bacteria purple.

During the same period, microscopes also improved dramatically. By the end of the 1880s, compound microscopes could magnify objects more than 2,000 times their actual size, with resolutions up to 1,000 times greater than that of the human eye.

A First Look at Chromosomes

As early as the 1840s, German scientists reported seeing strange threadlike objects inside the nuclei of some cells. As microscopes and staining techniques improved, other researchers made similar observations. Then, in the 1870s, a German scientist named Walther Flemming reached a critical milestone as he examined salamander cells. Using a dye developed in 1871, Flemming clearly saw that the threadlike objects were a standard feature of plant and animal cells. These structures are called chromosomes.

Flemming spent many hours observing chromosomes with a high-power microscope. He saw that just before a cell divides, its chromosomes line up in the middle of the cell. Then they break in half and move to opposite sides of the cell. One half of each original chromosome ends up in the two new cells. Flemming called the process mitosis. *Mitosis* comes from a Greek word for "thread."

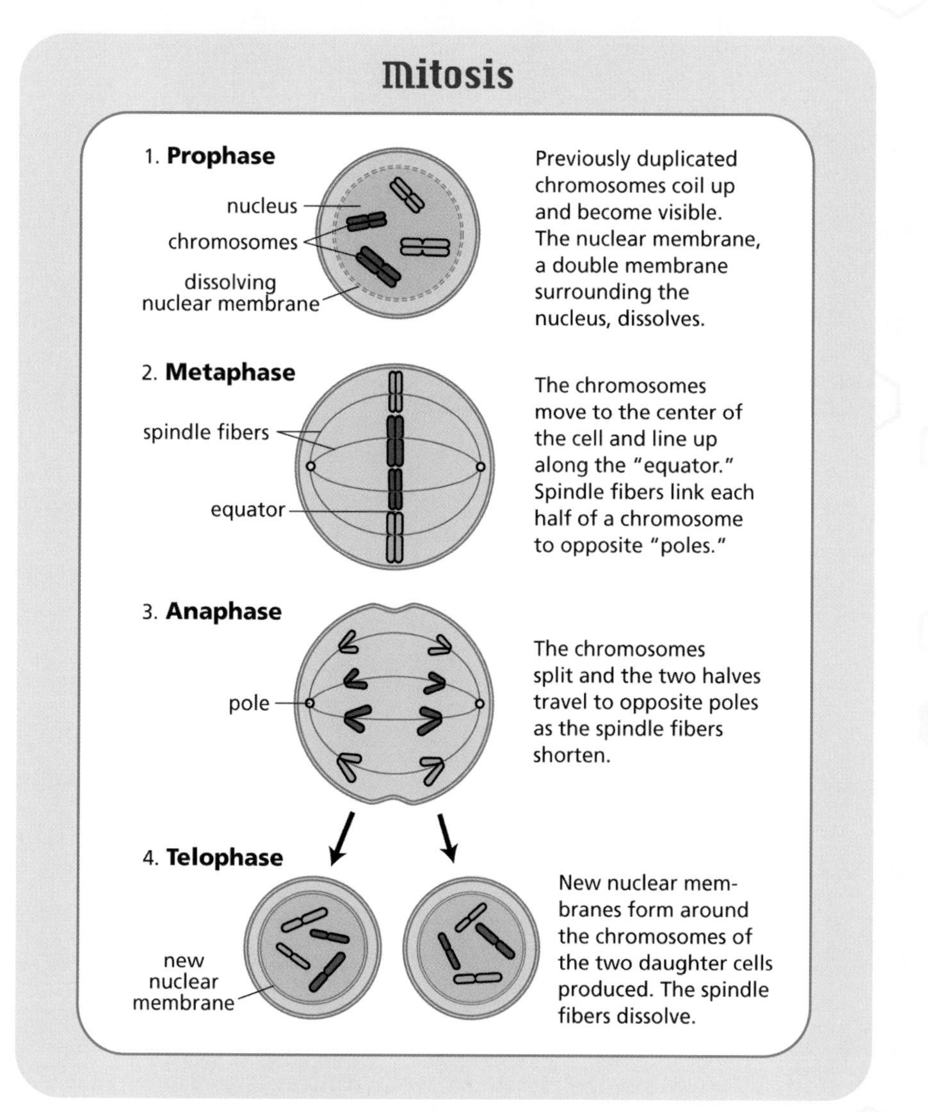

German botanist Eduard Strasburger was fascinated by Flemming's findings. Strasburger wondered if plant cells also undergo mitosis. When Strasburger stained plant cells with the same dye and looked at them under his microscope, he immediately had his answer. His observations matched Flemming's descriptions exactly.

As Strasburger continued to study mitosis, he noticed that the process seemed to occur in three distinct stages. He named these prophase, metaphase, and anaphase. Later scientists broke anaphase into two stages, calling the final stage of mitosis "telophase."

The Importance of Chromosomes

Strasburger wasn't the only researcher interested in chromosomes and mitosis. With new tools and techniques at their disposal, scientists began a flurry of research in this area. While Strasburger studied plants and Flemming worked with salamander cells, a Belgian scientist named Edouard van Beneden focused his attention on roundworms.

Beneden was interested in how roundworms reproduce. He decided to look at the chromosomes in their gametes—their egg cells and sperm cells. What Beneden saw surprised him. Gametes had only half as many chromosomes as other body cells.

One of the scientists paying close attention to Beneden's work was German researcher August Weismann. Early in his career, Weismann had spent hours studying the eggs of hydrozoa—ocean-dwelling relatives of jellyfish—under a microscope. But in the 1860s, he began to have vision problems. By the 1880s, Weismann was still proposing interesting theories, but he could no longer back up his ideas with microscopic observations.

After learning of Beneden's discovery, Weismann came up with a possible explanation. At the time, scientists knew that during fertilization, a new individual is created when an egg cell from a female joins with a sperm cell from a male. As a result, half the chromosomes in a fertilized egg come from its mother and half come from its father.

According to Weismann, if gametes contained the complete set of chromosomes found in other body cells, the embryo would have twice as many chromosomes as its parents. Imagine how this number would multiply with each new generation. Eventually there would be so many chromosomes that they wouldn't all fit inside the nucleus. To prevent that from happening, as gametes form, they undergo a process that reduces their chromosome number by half.

A few years later, a German scientist named Theodor Boveri confirmed Weismann's theory. Boveri studied cell division in sea urchins' testes (the organs that produce sperm). He found that gamete formation involves two rounds of cell division. During the first division, the number of chromosomes in each new cell is reduced by half. The second division is very similar to mitosis. This two-step process is called *meiosis*. The name comes from a Greek word that means "to lessen or reduce."

These results led scientists to suspect that chromosomes play an important role in heredity—the passing of key physical and behavioral traits from one generation to the next. Weismann proposed that all living

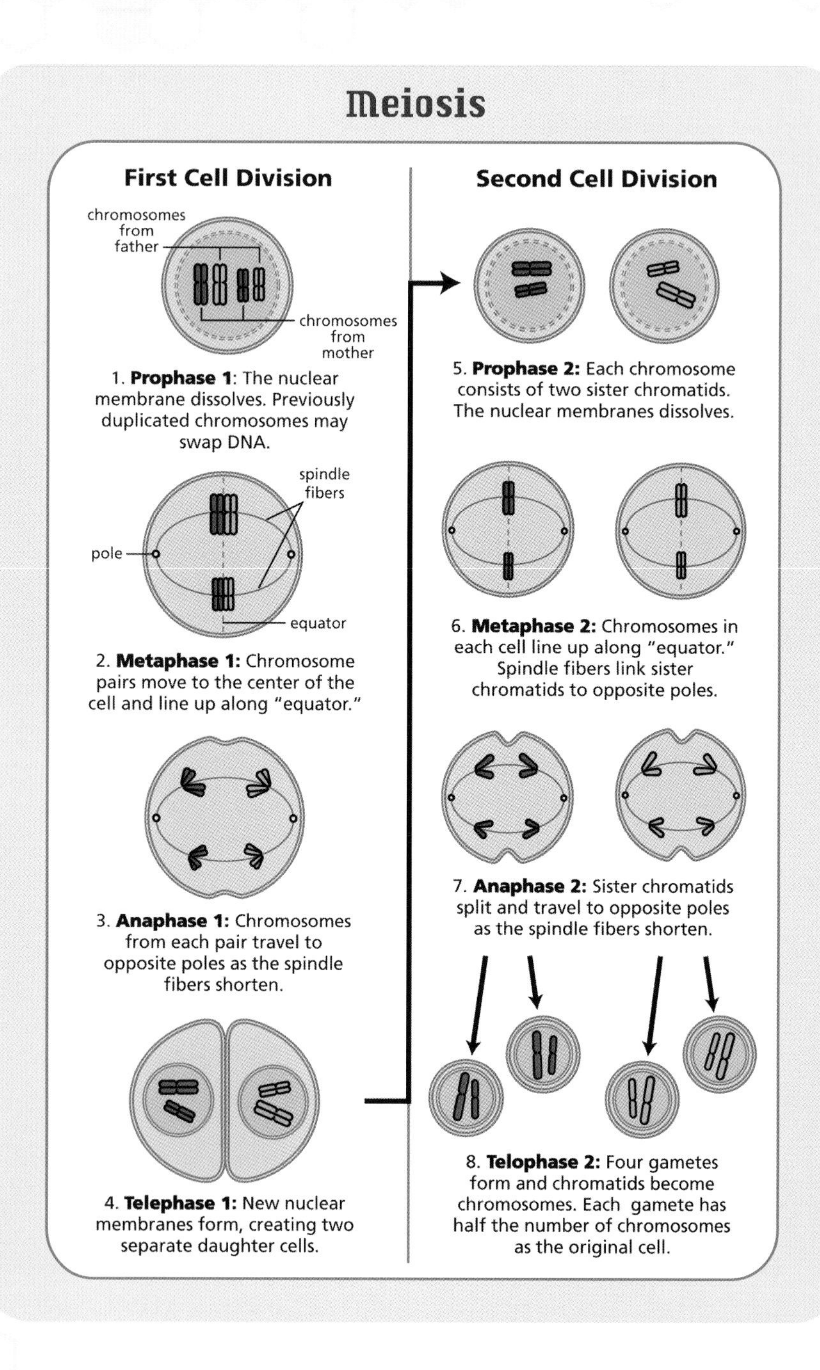
Meiosis
First Cell Division
chromosomes from father
chromosomes from mother
1. Prophase 1: The nuclear membrane dissolves. Previously duplicated chromosomes may swap DNA.
spindle fibers
pole
equator
2. Metaphase 1: Chromosome pairs move to the center of the cell and line up along "equator."
3. Anaphase 1: Chromosomes from each pair travel to opposite poles as the spindle fibers shorten.
4. Telephase 1: New nuclear membranes form, creating two separate daughter cells.
Second Cell Division
5. Prophase 2: Each chromosome consists of two sister chromatids. The nuclear membranes dissolves.
6. Metaphase 2: Chromosomes in each cell line up along "equator." Spindle fibers link sister chromatids to opposite poles.
7. Anaphase 2: Sister chromatids split and travel to opposite poles as the spindle fibers shorten.
8. Telophase 2: Four gametes form and chromatids become chromosomes. Each gamete has half the number of chromosomes as the original cell.

things contain a special hereditary substance. He believed that the substance controlled an organism's development and was passed from parent to offspring. Was this hereditary material located on chromosomes?

The Chromosomal Theory of Inheritance

By the end of the 1800s, cell biology had made extraordinary strides. To recap the advances, American scientist Edmund B. Wilson published *The Cell in Development and Heredity* in 1896. This influential book played an important role in shaping scientific understanding of the cell and its role in growth, development, and inherited traits.

Wilson spent most of his career teaching and doing research at Columbia University in New York City. Over the years, many exceptional graduate students worked in Wilson's lab. One of them was Walter Sutton.

Sutton made an important discovery even before he arrived at Columbia. In 1900, while studying grasshopper cells at the University of Kansas in Lawrence, Sutton noticed that chromosomes come in many different sizes and shapes. In body cells, each chromosome has an identical partner. But in gametes, chromosomes do not come in pairs. In other words, body cells have two complete sets of chromosomes, while gametes have just one complete set.

Sutton quickly grasped the importance of this observation. It meant that during meiosis, each chromosome pair separates, so that one complete set of chromosomes ends up in each gamete. It was not a random process. Instead, it was carefully orchestrated so that gametes only had half as many chromosomes as other cells.

Gregor Mendel

One of Sutton's first tasks in his new lab at Columbia was to write a report about his discovery. While he was doing this, a British scientist named William Bateson visited New York. Bateson had recently finished translating a scientific paper written in 1857 by an Austrian monk named Gregor Mendel. At the time, the paper received little attention. But in 1900, three scientists working independently all rediscovered Mendel's work. By this time, the scientific world was finally ready to appreciate Mendel's theories.

Working in a small garden at his monastery, Mendel bred generation after generation of pea plants. His work convinced him that certain essential physical traits, such as flower color, seed shape, and plant height, are passed from parent to offspring. Mendel developed three basic laws of inheritance. They could be used to predict how traits would be transmitted from one generation to the next by some unknown factors.

When Bateson first read Mendel's paper, he was studying inheritance in chickens. He was thrilled to discover that his own results fit perfectly with Mendel's ideas. Bateson realized that Mendel's laws of inheritance didn't apply to just peas. They applied to all plants and animals.

Bateson had come to New York in 1902 to spread the word about Mendel's work. It was probably at one of Bateson's lectures that Sutton first heard about the laws of inheritance. Not long afterward, Sutton finished his paper. He boldly stated that Mendel's "factors" were located on chromosomes. Instead of "factors," though, Sutton used the term "genes."

Meanwhile, Theodor Boveri was still studying sea urchin chromosomes in Germany. In one experiment, he noticed an egg that had been fertilized by two sperm—not just one. Because the egg received too many chromosomes, it didn't develop normally. Boveri realized that normal development depended on a cell getting just the right number of chromosomes. Thus, he reasoned, they must contain the hereditary material.

A few years later, Edmund Wilson brought together Boveri and Sutton's ideas to develop what he called the chromosomal theory of inheritance. This theory links Mendel's ideas with the activity scientists had observed during mitosis and meiosis. It also formed the cornerstone for a whole new branch of biology—genetics, or the study of genes and heredity.

By 1920 most scientists were convinced that chromosomes carry genes and that genes contain hereditary information that is passed from one generation to the next. But researchers still had one important question: what are genes made of?

DNA: The Hereditary Material

Scientists had known for a long time that cell nuclei contain proteins. In 1868 Friedrich Miescher, a Swiss doctor and biologist, had shown that nuclei also contain a molecule called deoxyribonucleic acid (DNA).

Miescher was very intelligent, but also very shy. Not long after earning his medical degree, he decided to focus on research rather than treating patients. One of his

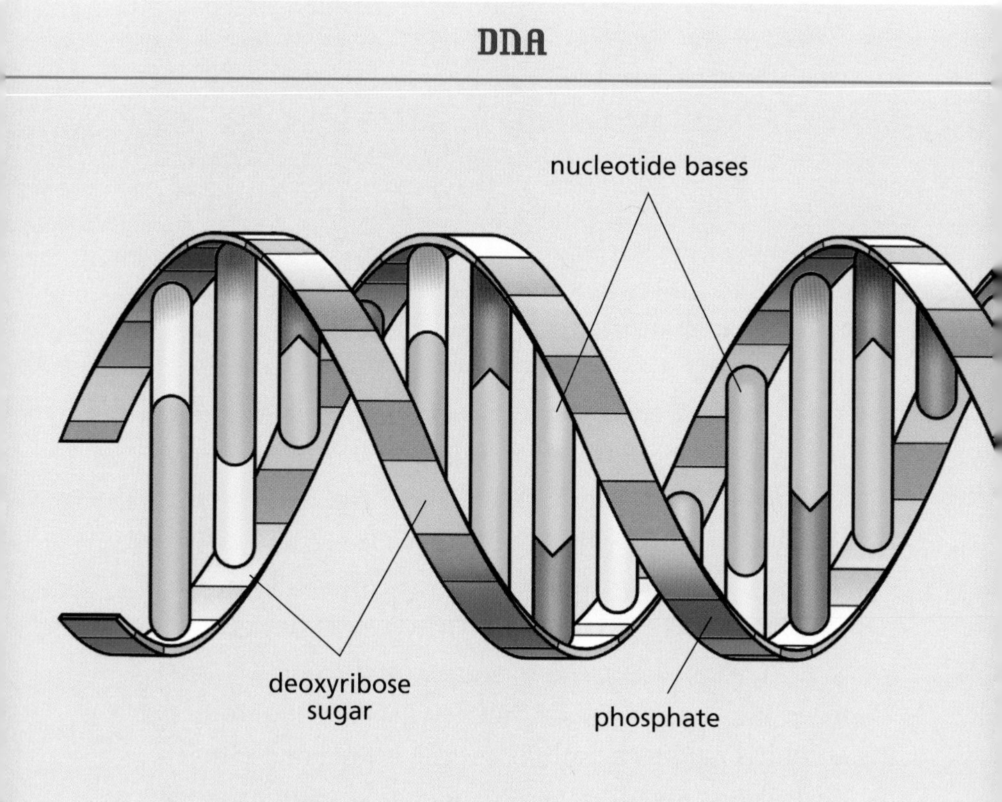

teachers suggested that he study the nuclei of white blood cells. Miescher agreed, even though he'd have to collect the cells from pus on used surgical bandages.

Miescher scraped the pus into glass containers and added a chemical that caused the cells to break open. The cells' heavy nuclei fell to the bottom of the container while other organelles rose to the top. Next Miescher extracted an unknown material from the bottom of the containers. He called it "nuclein" because it came from the blood cells' nuclei, but he wasn't sure what role it played in the cells.

As time passed, scientists learned that nuclein, later named DNA, contains phosphate, a sugar called deoxyribose, and four different chemical compounds called nucleotide bases. Researchers were intrigued by this strange new substance. But most scientists thought it was too simple to have the important job of transmitting hereditary information. They saw proteins as a more likely candidate.

Then, in the early 1940s, a team of researchers at the Rockefeller Institute for Medical Research (now Rockefeller University) in New York City proved that DNA, not a protein, is the hereditary material in bacteria. In 1952 scientists at Cold Spring Harbor Laboratory on Long Island, New York, showed that DNA is also the hereditary material in some viruses. Scientists realized that this was no coincidence. If DNA was the hereditary material in organisms as different as bacteria and viruses, it must also transmit genes in plants, animals, and all other living things.

Discovering DNA's Structure

Meanwhile, researchers in Europe were trying to understand the structure of DNA. They thought that if they could figure out what DNA looked like, they might be able to determine how it controlled cell functioning. Maybe they could even learn how it transferred key information from one generation to the next.

At King's College in London, England, a physicist named Maurice Wilkins and a graduate student, Raymond Gosling, were making images of DNA using X-ray diffraction. In this technique, a machine directs a tiny stream of X-rays at a sample. When the X-rays strike the sample, they scatter in all directions and create an image that shows the positions of the atoms and molecules that make up the sample.

In 1951 a brilliant young chemist named Rosalind Franklin came to King's College. She had been working with X-ray diffraction for several years and had invented new techniques for producing very detailed images of molecules. By 1952 she had produced some excellent images of DNA. She hoped to figure out the molecule's hidden secrets.

Without asking Franklin's permission, Wilkins showed some of her DNA images to two scientists working at Cavendish Laboratories in Cambridge, England. They were American biologist James Watson and British biophysicist Francis Crick.

Watson and Crick were trying to build a model of DNA. They had a solid understanding of the chemical compounds that made up the molecule, but they were struggling to figure out how all the pieces fit together.

When the two men saw Franklin's X-ray diffraction images, they realized that DNA is a double helix—a structure shaped like a spiraling ladder. It has sugar-phosphate supports on each side and rungs consisting of a bonded (linked) pair of nucleotides.

Watson and Crick rushed back to their lab and started working on their model. They knew that if they described the structure of DNA before anyone else, they would be famous. They might even win a Nobel Prize in Physiology or Medicine.

They were dismayed to learn that an American chemist named Linus Pauling had published a paper announcing the structure of DNA. When Watson and Crick

James Watson *(left)* and Francis Crick *(right)* with their model of a DNA molecule

read the paper, however, they breathed a sigh of relief. They knew Pauling was wrong. He hadn't taken into account some key pieces of information. Still, Pauling was a top-notch researcher. Watson and Crick knew it wouldn't be long before he discovered his mistake. They hurried to finish their model. Then they studied it carefully to see if they could find any flaws.

As the scientists looked at their completed model, they understood how the DNA molecule could easily copy itself each time a cell divided. The bonds between nucleotide bases could break and new strands would form from materials available inside the cell. The molecule that had been a mystery for so long now sat in front of the scientists, its secrets revealed.

No Nobel for Franklin

In 1962 Watson and Crick received the Nobel Prize for their work on DNA. They shared this honor with Maurice Wilkins, but Rosalind Franklin was not recognized.

Rosalind Franklin

Nobel Prizes are awarded only to living people. Sadly, Franklin had died of cancer in 1958. Many people believe that if she had lived, she would have received credit for her role in the discovery of the structure of DNA. Then she would have become the second woman (Gerty Theresa Cori was the first in 1947) to receive a Nobel Prize in Physiology or Medicine.

Chapter 5

Outside the Nucleus

While some researchers headed off to study DNA in greater detail, others realized that they still had plenty to learn about other parts of the cell. They turned their attention to the structure and function of the cell components that lay outside the nucleus. Thanks to devices and techniques developed in the early to mid-1900s, scientists could study cell organelles in a variety of new ways.

The first important new invention was the ultracentrifuge, a device that spins substances at high speeds. The spinning causes the substance's parts to separate based on their size and density. To use an ultracentrifuge, scientists place a sample, such as blood or other tissues, in test tubes. As the machine spins the test tubes, the heaviest components are forced to the bottom, while the lightest components remain on top.

Swedish scientist Theodor Svedberg designed the ultracentrifuge because he wanted to isolate proteins from tissue samples. Cell biologists soon realized they could use the device to separate organelles called mitochondria from the rest of the cell.

Scientists had known about mitochondria since 1857, when Swiss researcher Rudolf Kölliker first noticed the small, rod-shaped structures in muscle cells. For more than one hundred years, no one had known what mitochondria do. By using the ultracentrifuge, scientists could study the properties of mitochondria and determine their role in the cell. Svedberg received a Nobel Prize in Chemistry for his invention in 1926.

Cell biologists were thrilled that they could isolate mitochondria. But they also wanted to separate a cell's nucleus from its cytoplasm—the transparent, jellylike material between the cell's nucleus and its membrane. In 1938 German scientist Martin Behrens developed a new kind of centrifuge that could isolate the nucleus. During the 1940s and 1950s, scientists in the United States adapted Behrens's centrifuge to isolate a variety of other organelles.

A New Kind of Microscope

Because most organelles are too small to see with a light microscope, scientists did not discover them until an amazing new tool was developed by a German engineering student. As a young man, Ernst Ruska became interested in optics (the science of light). He knew that light microscopes could only create images of objects that were larger than the length of the light waves reflecting off them. As a result, a light microscope's magnifying power is limited to about 2,000 times.

Ruska also knew that electrons—tiny negatively charged particles inside atoms—have much shorter wavelengths than light rays do. He wondered if he could use a beam of electrons to create a microscope with enough magnifying power to view organelles and other extremely small objects. He decided to find out.

The transmission electron microscope (TEM) has been called one of the most important innovations of the twentieth century.

In the early 1930s, Ruska and his university adviser, Max Knoll, designed and built the world's first transmission electron microscope (TEM). By 1933 Ruska was building TEMs that could magnify objects ten times more than the best light microscopes.

Ruska received a Nobel Prize in Physics in 1986 for developing the TEM. By that time, it could magnify objects up to 1,000,000 times their actual size. Many scientists consider the electron microscope one of the most important innovations of the twentieth century.

Getting to Know Bacteria

Scientists quickly realized that Ruska's electron microscopes could revolutionize the study of cell biology. By 1937 a French marine biologist named Edouard Chatton was using an electron microscope to study a wide range of bacteria.

These single-celled creatures had received very little attention from scientists since Antoni van Leeuwenhoek first observed and described them in the 1670s. By the 1850s, researchers had learned a lot about plant cells, animal cells, and even protist cells, but they still knew almost nothing about bacteria.

Then, in 1857, French scientist Louis Pasteur published a scientific paper describing what he called the germ theory. Pasteur explained that when small microbes, which he called germs, got into vats of plant material that was being fermented to make wine or beer, the results were undrinkable. Instead of alcohol, beer brewers and winemakers ended up with sour-tasting acetic acid (vinegar).

Pasteur took his theory one step further. He suggested that germs might do more than ruin alcohol. They might also cause contagious diseases. Most scientists and doctors considered Pasteur's theory ridiculous. They couldn't imagine tiny microorganisms killing creatures as large as people or farm animals. But Pasteur didn't listen to the criticism. Instead, he began looking for ways to prove that germs carry diseases from one person or farm animal to another. Eventually he succeeded.

Because Pasteur contributed so much to modern knowledge of disease-causing microbes such as bacteria, many people think of him as the father of microbiology, the branch of biology that deals with microscopic forms of life. His work also led to the modern fields of bacteriology (the study of bacteria) and immunology (the study of how the body's immune system fights diseases).

By the 1880s, scientists understood the role bacteria play in many diseases, but they still knew almost nothing about the cells themselves. Because bacteria are so small, even the most powerful light microscopes can't reveal the details of their structure. That's why Edouard Chatton was so intent on studying bacteria with an electron microscope.

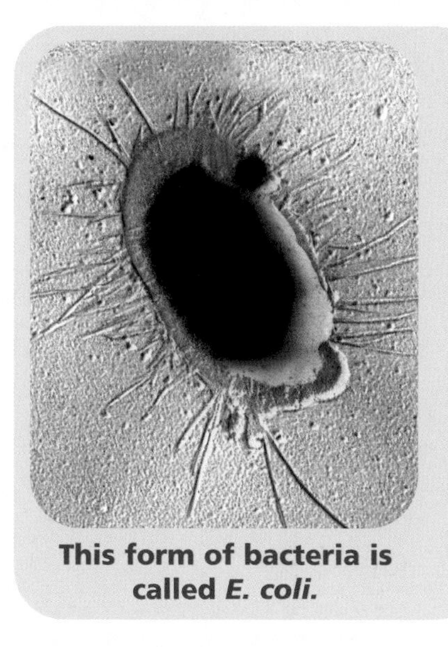

This form of bacteria is called *E. coli.*

Chatton's observations showed that bacteria cells were very different from plant and animal cells. Bacteria are made of just one cell, with no nucleus, no chromosomes, and only one kind of organelle. Their DNA floats freely in the cell, and they reproduce by simply splitting in half.

Chatton proposed that living organisms should be divided into two separate groups, the prokaryotes and the eukaryotes. The prokaryote ("before nucleus") group includes all bacteria. The eukaryote ("true nucleus") group includes plants, animals, fungi, and protists. Eukaryotes are larger and more complex than prokaryotes. Some are composed of one cell, while others contain many cells. A eukaryotic cell has a nucleus that contains DNA located on chromosomes. Eukaryotes have several different kinds of organelles and a much more sophisticated way of reproducing.

The Cradle of Cell Biology

Around the same time, a group of energetic young scientists at the Rockefeller Institute began using electron microscopes to study animal cells. Over the next two decades, Rockefeller Institute came to be known as the "cradle of cell biology."

Their first discovery—a new organelle—was made by Belgian American scientist Albert Claude in 1942. Another Rockefeller scientist, Canadian American Keith Porter, eventually gave the organelle its name: the endoplasmic reticulum (ER). Claude determined that this organelle plays a role in making and transporting fats and proteins. He also found that mitochondria produce energy for the cell. Porter continued studying the endoplasmic reticulum. He discovered that some ERs have smooth surfaces and others have rough surfaces.

In 1945 Porter and Claude, along with American scientist Ernest Fullam, published the first electron micrograph of a cell. An electron micrograph is a photographic image of an object as it appears when viewed with an electron microscope. The cell in the first micrograph came from a developing chick and was magnified 1,600 times. The image clearly showed mitochondria, the endoplasmic reticulum, and a structure that Italian physician and scientist Camillo Golgi had first spotted in nerve cells in 1897.

Golgi reported his findings and named the new organelle after himself. But it was so small and difficult to see with a light microscope that many scientists claimed the structure did not exist. It took almost fifty years and the power of an electron microscope to prove Golgi was right. His organelle is still known as the Golgi apparatus or Golgi body.

In 1946 Romanian scientist George Palade joined the group at Rockefeller Institute. At first he focused on developing better ways of preparing tissues for viewing with the electron microscope. Working with Porter, he designed a new kind of microtome and looked for ways to improve centrifuge techniques.

Later, Palade observed a variety of organelles closely and determined their roles in the cell. He discovered that mitochondria produce a chemical called adenosine triphosphate (ATP), which provides energy to cells. He also proposed that proteins are produced in ribosomes, tiny organelles that cover the surface of rough endoplasmic reticulum. And he showed that the Golgi apparatus sorts and helps transport large molecules within the cell.

In 1949 Claude returned to Belgium. The following year, Porter took charge of the lab. In 1953 Palade was also promoted. For close to a decade, Porter and Palade nurtured a generation of talented cell biologists.

Porter completed studies showing that organelle structure is consistent throughout a wide range of organisms, including protists, plants, and animals. Meanwhile, Palade continued to study the endoplasmic reticulum, learning more about its role in protein production. Palade also showed how tiny sacs called vesicles can merge with a cell's outer membrane and move materials into and out of the cell.

Another New Organelle

In 1949 Belgian scientist Christian de Duve was studying the breakdown of sugar in the liver. Using centrifuge techniques pioneered by Albert Claude, de Duve's team separated the components of liver cells according to their size, shape, and density.

After isolating the nucleus and known organelles from the mixture, they were left with an unknown substance. Further testing revealed a new organelle: the lysosome. Lysosomes contain digestive enzymes that break down waste material. All cells contain some lysosomes, but liver cells contain very high numbers. That makes sense, since one of the liver's main jobs is to filter waste products from the blood.

Surrounding the Cell

In 1957 J. David Robertson, a researcher at Harvard University in Cambridge, Massachusetts, made an important breakthrough while studying the cell's outermost layer—the cell membrane—with a transmission electron microscope. Robertson realized that the membrane consisted of two layers of fatty molecules called lipids. This type of structure was first proposed by German scientist

Charles Ernest Overton in 1895, but he did not have a powerful enough microscope to prove his idea.

Even with an electron microscope, Robertson was not able to see that the cell membrane contains another important component—proteins. That discovery was made by American scientists Jonathan Singer and Garth Nicolson in 1972. Eventually scientists realized that the membranes surrounding cell organelles are also composed of lipids and proteins.

Each cell type and organelle has a different combination of proteins in its membrane. For example, muscle cells have a large number of protein channels. Like a doughnut, a protein channel has a hole in the center. Through this passageway, materials can easily enter or exit a cell. In muscle cells, protein channels allow nutrients to enter quickly so the cells have a constant supply of energy.

The membranes surrounding the cells in sensory organs, such as eyes and ears, have an extra supply of molecules called protein receptors. They convey chemical messages through the cell membrane quickly and efficiently. Protein markers embedded in the membrane of every body cell act like name tags. They identify the cell as part of you, so your immune system won't attack it.

Scanning Electron Microscopes

In 1965 a new kind of microscope, the scanning electron microscope (SEM), became available. The first SEM had been built in the late 1930s by a brilliant German physicist named Manfred von Ardenne. But when World War II broke out in 1939, Germany's Nazi government forced Ardenne to switch his research focus to nuclear science. After the war, Ardenne was taken into custody by the Soviet Union and put to work developing atomic bombs. He was never able to continue his work on SEMs.

But in 1948, British physicist Charles Oatley began his own SEM research program at Cambridge University. With the help of nearly a dozen graduate students, Oatley designed and built a working SEM by 1961. Then he arranged for the Cambridge Instrument Company to begin producing the new microscope. The first five commercial SEMs were completed in 1965.

Scientists quickly discovered the advantages of the new device. Transmission electron microscopes can only

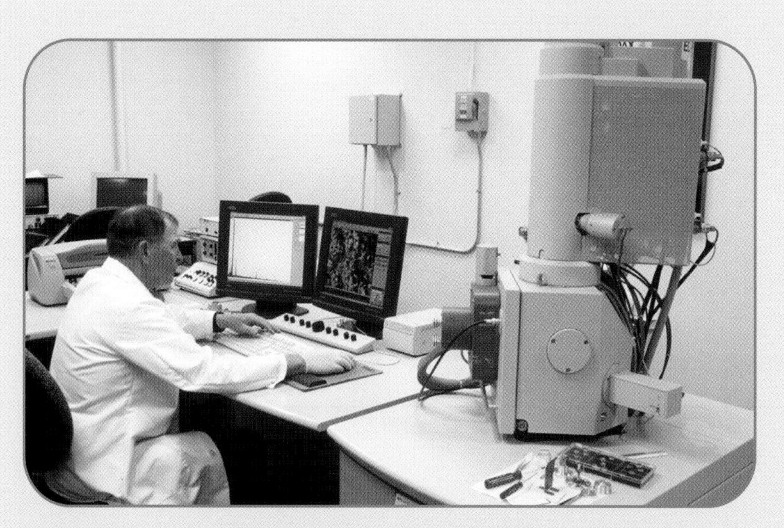

A researcher uses a scanning electron microscope (SEM), which can study whole samples of objects instead of slices.

How Electron Microscopes Work

When the metal filament (thin wire) at the top of an electron microscope heats up, a beam of electrons flows straight down a long tube. Magnets placed along the edges of the tube bend the electron beam, so that it can create a magnified image of a specimen and bring it into sharp focus.

In a TEM, specimens must be thinly sliced. When the electrons strike a specimen, some are absorbed. Others bounce off the object. But most pass through and form a magnified image. TEMs can magnify an object up to one million times.

Before a specimen can be viewed with a SEM, the object must be dried and coated with a thin layer of metal, such as gold. This makes most of the electrons bounce off the object rather than pass through it. As the electron beam scans, or moves across, the whole sample, the scattered electrons create a magnified, three-dimensional view of the object's surface. SEMs can only magnify objects about 100,000 times, but they allow scientists to examine specimens without cutting them up.

examine thinly sliced samples of a material. But the SEM can be used to study whole samples. With this new tool, scientists could see the surfaces of cells and organelles.

Researcher Keith Porter immediately saw the value of SEMs for studying cells. In 1968 he moved to the University of Colorado at Boulder and convinced the school to buy a SEM. Within a few years, he and his students had developed techniques for viewing the surfaces of tissues, cells, and cell components in greater detail than ever before.

CHAPTER 6

THE CELL AT WORK

Since the day in 1665 when Robert Hooke first saw cells, our knowledge about and understanding of these tiny structures has grown tremendously. As light microscopes and staining techniques improved in the early 1800s, scientists observed first plant cells and then animal cells in greater detail. By the middle of that century, scientists had established that cells are the basic units of all living things.

In the late 1800s and early 1900s, cell biologists focused on the nucleus. During this period, they identified chromosomes and observed cell division. In the mid-1900s, electron microscopes allowed researchers to get an up-close view of bacteria as well as the tiny organelles at work inside plant and animal cells.

By the 1970s, scientists had a solid understanding of both the structure and function of cells, including the specific roles of organelles. Each organelle is like a separate workstation or assembly line in a complex biochemical factory. It works by itself but contributes to the whole. In 1974 the Nobel committee recognized the tremendous advances in cell biology by awarding the prize for Physiology or Medicine to three men who had made important contributions to the field: Albert Claude, George Palade, and Christian de Duve.

As our knowledge of cells has increased, so has the desire to know even more. Every answer fuels a new set of questions. Each discovery leads to new possibilities for understanding the mysteries of life—and the secrets of death. That's why more people than ever are studying the cell. Their hard work and dedication will spur future advances.

Directing the Cell

Scientists have known for many years that nearly all plant and animal cells have a nucleus that contains chromosomes. More recently, they discovered that each chromosome consists of a long DNA strand wrapped around proteins. The nucleotide bases in DNA carry instructions for building proteins. Proteins are the workhorses of every cell. Each protein is designed to carry out a specific job within the cell. Without proteins, cells could not function.

Messages carrying DNA's genetic code pass out of the nucleus into the cell's jellylike cytoplasm and travel to the rough endoplasmic reticulum. This organelle has a very large membrane with many branching tubes and flattened sacs. Its rough appearance comes from the ribosomes covering it. Ribosomes are the site of protein synthesis. During this process, proteins are constructed from amino acids floating in the cytoplasm. The amino acids are linked together into a long chain to make the protein.

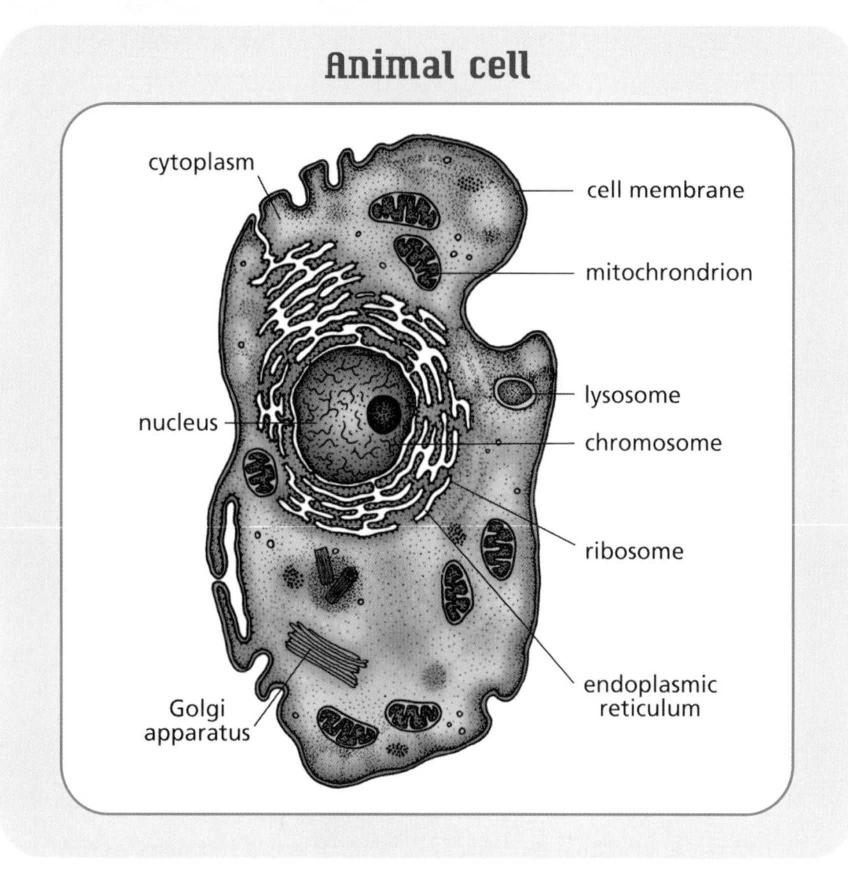

Animal cell

When the proteins are complete, they travel to the Golgi apparatus. This organelle looks like a stack of deflated balloons. It labels the proteins with tags that specify their destination. Some travel to other organelles, while others become part of the cell membrane. A typical cell can produce about 30,000 different kinds of proteins. Some are used to build or repair parts of the cell. Proteins called enzymes make chemical reactions occur more quickly or at a lower temperature.

Some endoplasmic reticula, known as smooth ERs, lack ribosomes. They produce the lipids that make up the cell membrane and the membranes surrounding some of the organelles. Smooth ERs also produce and store carbohydrates, and they manufacture lysosomes, the small organelles discovered by Christian de Duve.

Lysosomes act as the cell's recycling center and garbage disposal. Powerful digestive enzymes inside lysosomes break down worn-out organelles and ship their building blocks to the cytoplasm. There they are used to build new organelles. Lysosomes also dismantle and recycle proteins, lipids, and other molecules.

Powering the Cell

Cells get the energy they need to do their jobs from mitochondria. In a complicated set of chemical reactions know as respiration, nutrients are converted into adenosine triphosphate (ATP) molecules. ATP is the main source of energy for cells. In a typical cell, mitochondria produce hundreds of thousands of ATP molecules every minute. As the energy in ATP is used up by the cell's activities, ATP loses a phosphate molecule and is converted into adenosine diphosphate (ADP). ADP is returned to the mitochondria and used to build more ATP.

In animals, nutrients come from food particles broken down by the digestive system. But in plants, the nutrients are the end product of photosynthesis. This important process occurs inside chloroplasts. It could not occur without chlorophyll, which captures energy from the sun.

How Plant Cells Are Unique

Chloroplasts are not the only structure found in plants but not in animals. Plant cells also include a large vacuole, a central cavity that pushes all the other organelles to the edges of the cell. The vacuole stores water and many other materials. When a plant gets plenty of water, its vacuoles stay full, cell membranes are stretched tight, and the plant stands tall. But if too much water drains out of the vacuoles inside the plant's cells, it wilts. Vacuoles also contain pigments that give flowers their color. In some plants, vacuoles store bad-tasting substances that protect the plant from hungry animals.

An animal cell is surrounded and supported by a cell membrane. So is every plant cell. But plant cells also have a sturdy cell wall made of a protein called cellulose. The cell wall surrounds and protects the cell membrane and helps control the flow of water in and out of the central vacuole.

Studying Cells into the Future

In recent years, cell research has become more and more narrowly focused. Scientists no longer study plant cells, animal cells, or bacteria as a whole. Instead, they concentrate on particular cell parts or processes. As a result, the field of cell biology has branched out in many new directions.

Thus, even though many scientists study cells, these researchers do not call themselves cell biologists. Developmental biologists focus on the formation and development of egg cells. Molecular geneticists determine the order of nucleotides in DNA strands and map the location of genes on chromosomes. Evolutionary microbiologists are interested in the single-celled creatures that lived on Earth millions of years ago.

Many medical researchers also study cells. Some struggle to understand cancer cells or stem cells so they can fight diseases and save lives. Others focus on why cells age or how cells communicate with one another. They hope their work will help people live longer, healthier lives.

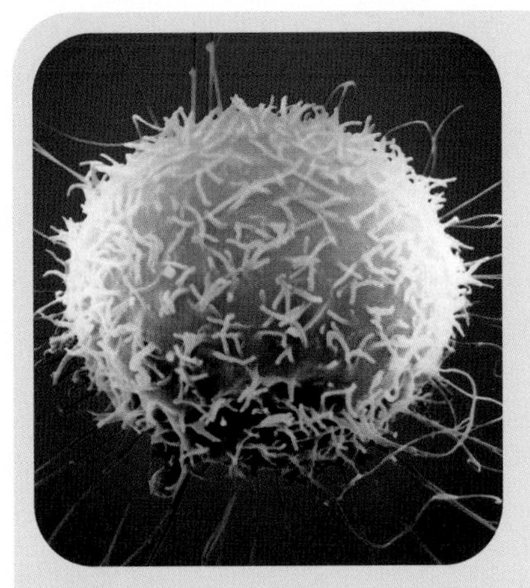

Many scientists study cancer cells to try to help fight disease.

What this means is that cell biology is no longer considered a separate branch of science. It has become an integral part of nearly every field of biology. The cell is that important.

In the future, scientists working in many different areas will continue to study cells. No one can predict all the amazing things they will learn. But their research will increase our understanding of humans and the many other creatures that share our world.

Glossary

bacterium (plural, bacteria): a tiny, one-celled organism without a nucleus

cell theory: the idea that all living things are made of one or more cells; that cells are the basic structural unit of life; and that all cells arise only from preexisting cells

chlorophyll: a pigment in plant cells' chloroplasts that captures energy from the sun's rays

chloroplast: an organelle in plant cells that collects the sunlight plants need to make their own food

chromosome: a rodlike structure that contains DNA and is located in the nucleus of most cells. Humans have twenty-three pairs of chromosomes.

concentration: amount of a fluid or other material in a given area

cytoplasm: the clear, jellylike material that fills most of the cell. Organelles are suspended in the cytoplasm.

density: the mass of an object in a given area or volume of space

diffusion: the movement of a substance from a place of high concentration to one of lower concentration

DNA (deoxyribonucleic acid): the molecule that contains genetic information and is passed from parent to offspring during reproduction. Shaped like a spiraling ladder, DNA is found in the nucleus of most cells.

embryo: an organism in its earliest stages of development

endoplasmic reticulum (ER): an organelle that moves proteins and other materials through the cell

eukaryote: an organism whose cell or cells have nuclei. Plants, animals, fungi, and protists are eukaryotes.

fertilization: the union of an egg cell and a sperm cell to create a new organism

gamete: an egg cell or sperm cell; a cell that takes part in fertilization

genes: the basic units of heredity, consisting of DNA

Golgi apparatus: an organelle that labels molecules made on the endoplasmic reticulum with chemical tags that specify their destination within or outside the cell

heredity: the passing of characteristics from one generation to the next

lipid: a key component of the membrane surrounding most cells

lysosome: an organelle that contains digestive enzymes that break down waste material

meiosis: division of sex cells to produce egg and sperm cells

mitochondrion (plural, mitochondria): an organelle that breaks down food molecules and produces the energy needed for all cellular activities

mitosis: the process by which the nucleus divides before cell division, producing two daughter cells with exactly the same number of chromosomes as the parent cell

nucleus (plural, nuclei): the control center of plant and animal cells, containing chromosomes with coded genetic instructions

organ: a group of tissues that works together, such as the stomach or heart

organelle: a structure within a cell that carries out a specific task

osmosis: the process by which water moves across a cell membrane from an area of higher concentration to one of lower concentration

photosynthesis: the process by which plants convert water and carbon dioxide into glucose (food) and water

prokaryote: a single-celled creature without a nucleus

protein: a large, complex substance that is made of amino acid chains. Proteins play many roles in cell functioning.

protist: a tiny organism with DNA enclosed in a nucleus. Protists are the ancestors of all plants, animals, and fungi.

ribosome: an organelle that serves as the site of protein production

stomata: tiny pores in plant leaves that open and close to take in carbon dioxide or get rid of oxygen and water

tissue: a group of cells that work together, such as nerve tissue or muscle tissue

ultracentrifuge: a device that spins samples, such as cells, at high speeds to separate the components

X-ray diffraction: a technique that can provide an image of the positions of atoms and molecules in a substance

Timeline

1595 Hans and Zacharias Janssen build the first compound microscope.

1665 Robert Hooke, who helped create a microscope with a built-in light source, describes what he calls "cells" in his book *Micrographia*.

1668 Francesco Redi proves that maggots do not form spontaneously.

1671 Marcello Malpighi and Nehemiah Grew describe their microscopic observations of cells.

1673 Antoni van Leeuwenhoek begins sending reports about his microscopic studies to the Royal Society of London.

1820s Henri Dutrochet and François-Vincent Raspail explain how materials move in and out of cells.

1833 Robert Brown describes the cell nucleus in orchid cells.

1838–1839 Matthias Schleiden and Theodor Schwann publish their cell theory.

1857 Rudolf Kölliker discovers mitochondria.

1858 Rudolph Virchow proposes that all cells come from preexisting cells.

1865 Julius von Sachs shows that chlorophyll is located in chloroplasts.

1868 Friedrich Miescher isolates DNA from cell nuclei.

1879 Walther Flemming documents chromosome behavior during mitosis.

1887 Theodor Boveri discovers that chromosomes contain hereditary material.

1896 Edmund B. Wilson publishes *The Cell in Development and Heredity*, an influential overview of cell biology.

1898 Camillo Golgi describes the Golgi apparatus.

1902 Walter Sutton discovers that chromosomes exist in pairs.

1926 Theodor Svedberg develops the ultracentrifuge.

1938 Martin Behrens develops a centrifuge that can separate nuclei from cytoplasm.

1939 Ernst Ruska helps Siemens Company produce the first electron microscope.

1945 Keith Porter, Albert Claude, and Ernest Fullam publish the first electron micrograph of a cell in *The Journal of Experimental Medicine.*

1953 James Watson and Frances Crick determine the structure of DNA.

1972 Jonathan Singer and Garth Nicolson describe the structure of the cell membrane.

1974 Albert Claude, George Palade, and Christian de Duve receive the Nobel Prize in Physiology or Medicine for their contributions to knowledge of cell biology.

1990 The Human Genome Project is launched to decode all the genes in human DNA.

1998 Researchers at the University of Wisconsin and Johns Hopkins University develop a method for isolating stem cells, which can develop into any kind of tissue.

2003 The Human Genome Project releases the final draft on the DNA sequence for the entire human genome—about 20,000 to 25,000 genes comprising nearly 3 billion nucleotides.

Biographies

Albert Claude (1899–1983) Albert Claude was born in Longlier, Belgium. After receiving a medical degree from the University of Liège in 1928, he moved to the United States and began working at the Rockefeller Institute for Medical Research. When electron microscopes became available in the early 1940s, Claude began to study the structure and function of cells. His pioneering work in this area earned him a Nobel Prize in 1976. Claude also developed an important method for separating cell components using a centrifuge. In 1948 Claude returned to Belgium and continued his research at the Free University of Brussels and the Jules Bordet Institute in Brussels. In 1971 Claude joined the Catholic University of Leuven, where he worked until his death in 1983.

Francis Crick (1916–2004) From an early age, Francis Crick showed an avid interest in science. His father, who owned a shoe factory in Northampton, England, encouraged the boy's love of experiments. Crick studied physics at University College in London and began work on his Ph.D. During World War II (1939–1945), he left the university to design mines for the British government. In 1947 Crick returned to his studies, focusing on biology. At first Crick studied proteins. But in 1951, he and James Watson began to work on the structure of DNA. Despite much competition, Watson and Crick were the first to understand the molecule's structure. In 1961 Crick and fellow biologist Sydney Brenner proved that the information in DNA is read in a series of three nucleotide bases. The following year, Crick shared the Nobel Prize with Watson and Maurice Wilkins. In 1976 Crick moved to the Salk Institute for Biological Studies in California, where he conducted research on neurobiology (biology of the nervous system). He died in 2004 at the age of eighty-eight.

Walther Flemming (1843–1905) Trained as a doctor in his native Germany, Walther Flemming went on to do research in the Netherlands, the Czech Republic, and finally back in Germany. Flemming was interested in cell-staining techniques. He developed a method that allowed him to identify chromosomes inside the nuclei of salamander cells. He was also the first scientist to describe how chromosomes behave during cell division. Flemming observed that chromosomes double during cell division and are then equally distributed to the two resulting daughter cells. He coined the term *mitosis* for this process. Flemming published his findings in 1882, but scientists did not understand the importance of his work until twenty years later.

Robert Hooke (1635–1703) After graduating from the University of Oxford, England, Robert Hooke assisted well-known physicist Robert Boyle. In 1662 he became the curator of experiments of the Royal Society of London, and in 1665 he was hired as a geometry professor at Oxford. Following the Great Fire of London in 1666, he became surveyor of the city. He designed many buildings, some of which are still standing. Hooke had many varied interests. His inventions include the iris diaphragm in cameras, the universal joint used in automobiles, and the balance wheel in watches. Working with instrument maker Christopher Cock, Hooke developed the first microscope with a built-in light source. In 1665 he published *Micrographia*, which includes descriptions and drawings of his observations.

Antoni van Leeuwenhoek (1632–1723) A native of Delft, Holland, Antoni van Leeuwenhoek made his living as a cloth merchant and a sheriff's deputy. But he is best known for his hobby—making and using microscopes. He was the first person to describe protists and bacteria. He also studied a variety of plant and animal tissues. Among other observations, he saw red blood cells flowing through the capillaries in a rabbit's ears and a frog's foot. Leeuwenhoek never taught anyone the secret behind his high-power microscopes, but he did write long letters describing his observations to the Royal Society of London.

Matthias Schleiden (1804–1881) Born in Hamburg, Germany, Matthias Schleiden was trained as a lawyer but decided to become a botanist. In the late 1830s, he observed plants with a microscope and realized that they are made of cells. Schleiden published his findings in 1838. A year later, his friend Theodor Schwann recognized that animals are also made of cells, and the two men developed the cell theory. Schleiden taught botany at the University of Jena, Germany, from 1839 to 1862. By the 1850s, his interests had shifted to philosophy and history. In his later years, he worked as a private tutor.

Theodor Schwann (1810–1882) Theodor Schwann was born in Neuss, Germany. As a boy, he enjoyed tinkering with mechanical devices. He studied medicine at the universities of Bonn, Würzburg, and Berlin. After graduating in 1834, he worked at an anatomy museum in Berlin. During this period, he discovered the digestive enzyme pepsin. He also studied fermentation and muscle movement. In 1838 Schwann began teaching anatomy at the University of Leuven in Belgium. After applying Matthias Schleiden's cell theory to animals in 1839, he showed that all mature animal tissue forms from embryonic cells. In 1848 Schwann moved to the University of Liège in Belgium and taught anatomy and physiology. He continued to study cells until his death in 1882.

Walter Sutton (1877–1916) Born in Utica, New York, Walter Sutton showed an early interest in mechanics and began studying engineering at the University of Kansas in 1894. After spending the summer of 1897 caring for sick family members, Sutton switched his major to biology. As a graduate student at the University of Kansas, Sutton noticed that chromosomes come in many different sizes and shapes and usually in pairs. The finding was so important that Sutton won a scholarship to continue his studies at Columbia University in New York. In 1903 he suddenly decided to return to Kansas and work in the oil fields. He invented several devices that made it easier and cheaper to extract oil from the ground. In 1905 Sutton returned to Columbia University and earned a degree in medicine. In 1909 he moved back to Kansas and became an assistant professor of surgery at the University of Kansas School of Medicine. During World War I (1914–1918), he spent several months working at a military hospital in France. In the summer of 1915, he returned to Kansas, where he continued to teach, publish, and perform surgeries. About a year later, Sutton died tragically following an operation for appendicitis. He was only thirty-nine.

RUDOLF VIRCHOW (1821–1902) Born in Poland, Rudolf Virchow was educated at the University of Berlin. His first job was dissecting corpses at a hospital in Germany. In 1847 he became a lecturer and then a full professor at the University of Würzburg. In 1856 Virchow returned to the University of Berlin. He contributed to cell theory by asserting that all cells arise from preexisting cells. He also showed that cell theory applies to diseased tissues as well as healthy ones.

JAMES D. WATSON (1928–) As a child, James Watson enjoyed memorizing unusual facts. This came in handy when he appeared on *Quiz Kid*, a popular radio program. He used his $100 prize to buy binoculars for bird-watching. He planned to become an ornithologist (a scientist who studies birds) until he read a book about genes and chromosomes. Then he decided to study genetics instead. In 1951 Watson met Maurice Wilkins and became interested in DNA. Two years later, at the University of Cambridge, Watson and graduate student Francis Crick figured out the structure of DNA and built a large model of the molecule. In 1962 Watson, Crick, and Wilkins shared the Nobel Prize for their work on DNA. A few years later, Watson accepted a position at Harvard University, where he studied ribonucleic acid (RNA). In 1968 he became director of Cold Spring Harbor Laboratory on Long Island, New York. From 1988 to 1992, Watson ran the Human Genome Project at the National Institutes of Health. In 1994 he returned to Cold Spring Harbor Laboratory and became the institution's president.

Source Notes

16 Antony van Leeuwenhoek, Letters to the Royal Society of London, September 17, 1683, *Antony van Leeuwenhoek*, N.d., University of California Museum of Paleontology, www.ucmp.berkeley.edu/ history/ leeuwenhoek.html (August 29, 2006).

16 Ibid.

30 Theodor Schwann, "Microscopical Researches into the Accordance in the Structure and Growth of Animals and Plants," *The Incredible Life and Times of Biological Cells*, N.d., http://fig.cox.miami.edu/~ddiresta/bil101/Cells.htm (August 29, 2006).

31 Ibid.

33 Rudolf Virchow, "Cellular Pathology," *Landmarks in the Study of Cell Biology*, N.d., http://fig.cox.miami.edu/ ~cmallery/ 150/unity/cell.text.htm (June 5, 2006).

54 "Keith Roberts Porter: 1912–1997." *Journal of Cell Biology*, July 28, 1997, 223–224.

Selected Bibliography

Abbott, David, ed. *The Biographic Dictionary of Scientists: Biology*. New York: Peter Bendick Books, 1984.

Cell Biology. Danbury, CT: Grolier Educational Publishing, 2004.

Ellavich, Marie C. *Scientists: Their Lives and Work*. Detroit, MI: UXL/Gale, 1999.

"The Facets of Cell Biology." *American Society for Cell Biology*. N.d. http:www.ascb.org/information/facets.htm.

Harris, Henry. *The Birth of the Cell*. New Haven, CT: Yale University Press, 1999.

Johnson, George B., and Gary J. Brusca. *Biology: Visualizing Life*. Austin, TX: Holt, Rinehart, and Winston, 1994.

Johnson, George B., and Peter H. Raven. *Biology: Principles and Explorations*. Austin, TX: Holt, Rinehart, and Winston, 2001.

"Landmarks in the Study of Cell Biology." N.d. http://fig.cox.miami.edu/~cmallery/ 150/unity/cell.text.htm.

"The Search for DNA: The Birth of Molecular Biology." *Access Excellence @ the national health museum*, 1999. http://www.accessexcellence.org/RC/AB/BC/Search_for_DNA.html.

Sheeler, Phillip, and Donald E. Bianchi. *Cell Biology: Structure, Biochemistry and Function*. New York: Wiley, 1983.

Further Reading

Fridell, Ron *Decoding Life: Unraveling the Mysteries of the Genome*. Minneapolis: Twenty-First Century Books, 2005.

Johnson, Rebecca L. *Genetics*. Minneapolis: Twenty-First Century Books, 2006.

Snedden, Robert. *Microlife: Scientists and Discoveries*. Chicago: Heinemann, 2000.

———. *The World of the Cell: Life on a Small Scale*. Chicago: Heinemann, 2003.

Stewart, Gail. *Microscopes: Bringing the Unseen World into Focus*. San Diego, CA: Lucent Books, 1992.

Stille, Darlene. *Extraordinary Women Scientists*. Danbury, CT: Children's Press, 1995.

Watson, James D. *The Double Helix: A Personal Account of the Discovery of DNA*. New York: Touchstone, 1968.

Yount, Lisa. *Antoni van Leeuwenhoek: First to See Microscopic Life*. Berkeley Heights, NJ: Enslow, 1996.

Websites

DNA from the Beginning
http://www.dnaftb.org/dnaftb/
This animated website lets you explore the history of genetics in a delightfully fun yet informative way.

How Cells Divide
http://www.pbs.org/wgbh/nova/baby/divi_flash.html
View clear, step-by-step diagrams of mitosis and meiosis at this site.

The Virtual Electron Microscope
http://school.discovery.com/lessonplans/activities/electronmicroscope/
Find out how microscopes have contributed to our knowledge about cells.

INDEX

Photo Acknowledgments

The images in this book are used with the permission of:
© age fotostock/SuperStock, p. 5; © Laura Westlund/Independent Picture Service, pp. 6, 7, 22, 23, 37, 40, 44, 62; © Science VU/Visuals Unlimited, p. 8; The Granger Collection, New York, p. 11; © Hulton Archive/Getty Images, pp. 12, 29 (left); © Charles D. Winters/Photo Researchers, Inc., p. 14; Courtesy of the National Library of Medicine, pp. 15, 42; © Michael Abbey/Visuals Unlimited, p. 16; © Henry Aldrich/Visuals Unlimited, p. 17; Library of Congress, p. 25 (LC-USZ62-72127); © Kean Collection/Hulton Archive/Getty Images, p. 29 (right); © Professor P.M. Motta & D. Palermo/Photo Researchers, Inc., p. 33; © A. Barrington Brown/Photo Researchers, Inc., p. 47; © Science Source/Photo Researchers, Inc., p. 48; © Dwight R. Kuhn, p. 51; © George Musil/Visuals Unlimited, p. 53; © Inga Spence/Visuals Unlimited, p. 58; © Dr. David M. Phillips/Visuals Unlimited, p. 65.

Front Cover: © Tim Parlin/Independent Picture Service.